MEMORIZING MULTIPLICATION

A Relaxed Approach to Math and the Multiplication Times Tables

Tony Guerra, Pharm.D.
2024

Copyright © 2024 by Tony Guerra, Pharm.D.

All rights reserved. No part of this book may be reproduced or transmitted in any form or by any means without written permission of the author.

Memorizing Multiplication: A Relaxed Approach to Math and Multiplication Times Tables

*To Mindy,
Brielle, Rianne, and Teagan*

TABLE OF CONTENTS

AUTHOR'S NOTE 1
  What is a Relaxed Approach? 1
  Why is multiplication hard? 3
  Why does it take so long to learn? 5
  Why does everyone else seem to know the times table? 6
  Why won't your brain just remember the times table? 6
  Why did I write this book as an audiobook? 8
  Math and the Dishwashing Machine 10
Introduction – 0s, All Zeros 11
Chapter 1 – 1s, Same Number 12
Chapter 2 – 2s, Add itself 13
Chapter 3 – 3s, Skip Counting and More 14
  Triplet A – 3s, Skip Counting – 1 to 4 15
  Triplet B –3s, Skip Counting – 5 to 8 15
  Triplet C – 3s, Subtraction, Adding Zero, Twins, Times 10, times 2 – 9 to 12 17
Chapter 4 – 4s, Skip Counting and More 20
  Triplet A – Skip Counting 4s – 1 to 4 21
  Triplet B – Skip Counting 4s – 5 to 8 22
  Triplet C – 4s, Subtraction, Adding Zero, Twins, Times 10, times 2 – 9 to 12 23
Chapter 5 – 5s, Skip Counting and More 25
  Triplet A – Skip Counting 5s – 1 to 4 26
  Triplet B – Skip Counting 5s – 5 to 8 26

i

Triplet C – 5s, Subtraction, Adding Zero, Twins, Times 10,
times 2 – 9 to 12 27

Chapter 6 – 6s, Double 3s and Rhymes 29
Triplet A – 6s, Double 3s and Rhymes – 1 to 4 29
Triplet B – 6s, Double 3s or Rhymes – 5 to 8 30
Triplet C – 6s, Subtraction, Adding Zero, Twins, Times 10,
times 2 – 9 to 12 30

Chapter 7 – 7s, Skip Counting, Rhymes, and More 33
Triplet A – 7s, Skip Counting – 1 to 4 33
Triplet B – 7s, Anchor and Rhymes – 5 to 8 33
Triplet C – 7s, Subtraction, Adding Zero, Twins, Times 10,
times 2 – 9 to 12 34

Chapter 8 – 8s, Double 4s. 36
Triplet A – 8s, Skip Counting and Double 4s – 1 to 4 36
Triplet B – 8s, Skip Counting and Double 4s – 5 to 8 37
Triplet C – 8s, Subtraction, Adding Zero, Twins, Times 10,
times 2 – 9 to 12 38

Chapter 9 – 9s, The Fingers Trick and More 40
Triplet A – 9s, Multiply by 10 and Subtract and the Fingers
Trick – 1 to 4 41
Triplet B – 9s, Multiply by 10 and Subtract and the Fingers
Trick – 5 to 8 41
Triplet C – 9s, Multiply by 10 and Subtract and the Fingers
Trick – 9 to 12 42

Chapter 10 – 10s, Add a Zero 44
Triplet A – 10s, Add a Zero – 1 to 4 44
Triplet B – 10s, Add a Zero – 5 to 8 44
Triplet C – 10s, Add a Zero – 9 to 12 44

threes, and learn three, six, nine, twelve and so on. You'll either do fours or move on to fives and tens which are a lot easier. But then, like a fun video game gone bad, the levels get super hard.

The sixes, sevens, eights, and nines seem to take forever. It takes so much longer than the earlier numbers, that many students quit and never finish learning the times table like I did when I was younger. But that's a real problem because if you don't know your times tables, you are going to struggle with division. And if you struggle with division, Algebra is a mess.

If you don't know multiplication and division off the top of your head, then a later grade problem like "If three x equals twelve, what is x?" (If 3 x = 12, What is x?) isn't something you can really do in your head like your classmates. You then slide further behind and get more frustrated.

By the way, the answer to "If three x equals twelve what is x" (If 3 x = 12, What is x?) is that "x equals four." (x = 4)

WHY DOES IT TAKE SO LONG TO LEARN?

I'm sure you've tried to write the times table many times and it just doesn't stick. Or maybe you've used a bunch of practice sheets or flash cards, and still, the information just slides off your brain like a good, gooey slime slides off a table.

What can do you do to make multiplication stick?

It takes two steps.

First, write out your strategy.

Instead of writing that two times nine equals eighteen (2 x 9 = 18), you would write down that "two times nine equals nine plus nine equals eighteen." (2 x 9 = 9 + 9 = 18) if your strategy is that for every number that is multiplied by two, you can add those two numbers together also.

Second, try teaching your methods to someone else.

Don't worry, you don't have to stand in front of a classroom like I do. You can teach it to your dog, cat, toy, little niece or nephew, someone you're babysitting and so on.

Your brain just cares that you are teaching it or really, more generally that you are sharing the technique. When you teach something, then your brain knows it's something important, and something your brain should remember.

WHY DOES EVERYONE ELSE SEEM TO KNOW THE TIMES TABLE?

It's a myth that everyone has the answer in their head. When you look at the one hundred forty-four (144) numbers of a twelve by twelve (12x12) times table from one (1) in the top left and one hundred forty-four (144) in the bottom right, there are only numbers there because the strategies don't fit in the table.

For example, not everyone knows what seven times nine equals (7 x 9 =) off the top of their head, a lot of people, including me either put their seventh (7th) finger, your right-hand pointer finger, down to see six fingers on the left and three fingers on the right to make six-three or sixty three. Or, alternatively, you can multiply seven times ten to make seventy and subtract seven to equal sixty-three. (7 x 10 = 70, - 7 = 63) In both cases, the answer didn't just pop into the head, rather, the small story of how you remembered it to share it did.

WHY WON'T YOUR BRAIN JUST REMEMBER THE TIMES TABLE?

You could draw the traditional times table in about fifteen minutes, and if your teacher let you, you could simply use that for every problem. But you can't use that times table to memorize the answers quickly because your brain doesn't remember numbers that way, it remembers stories.

Think about your dog when you talk to him. Our dog is Beau, B-e-a-u, like the first four letters in beautiful, b-e-a-u-t-i-f-u-l, because he's a handsome dog. When my dog hears you say words, it sounds like words, words, words, Beau, words, words, words, treat.

There is a neighbor's giant sheep dog that Beau plays with all the time named Gronk, G-r-o-n-k, and when he hears that word, he'll run through the sliding door to see if Gronk is waiting at the fence to play with him.

Your brain is the same, it hears, numbers times numbers equals answer. A different number times a different number equals another answer. Your brain just sees all the numbers the same unless there is a story or name with it, like Gronk.

Why is that important?

If you try to remember that seven times nine equals sixty-three (7 x 9 = 63), your brain might put that math fact in short term memory.

But then when you try to remember that six times seven equals forty-two (6 x 7 = 42), your brain remembers that in short term memory and kicks the seven times nine equals sixty three (7 x 9 = 63) out.

Again, your brain just hears number times number equals answer.

Just like you can teach a trick to your dog, you can trick your brain into remembering something. Our dog Beau will sit, shake, turn around, lie down, and wait for his treat. Not only does the dog remember what to do, you remember that the order of tricks is going to be sit, shake, turn around, then lie down.

If you teach the trick that seven times nine equals seven times ten minus seven equals sixty three (7 x 9 = 7 x 10 = 70 - 7 = 63) to someone else, maybe Beau, then your brain says, this is important enough to teach and help someone else, I better put this in my long-term memory, I might need this later.

So, each times table fact will have a little bit of a story, which gives the chewing gum its stickiness. At first, it might sound boring with the 1s and 2s and so on, but we're going to establish the technique with those easier numbers that will work for the harder numbers. Eventually, as you get the techniques, you'll remember the whole times tables.

It will be a little bit ironic, but eventually you will get so good that you will find it difficult to understand why it was so hard to learn in the first place – but that's a good thing.

WHY DID I WRITE THIS BOOK AS AN AUDIOBOOK?

The reason I wrote this book is because if you have a math book in your hand about multiplication and you are in a higher grade, you might be a little embarrassed. No one knows what you are listening to in an audiobook.

I also struggle with print books, I can't pay attention and can only make it through a few pages before I get sleepy.

Whether I'm working out at the gym, running with Beau, or driving to my work to teach, I love to listen.

So, don't feel like you need to do anything at first, just listen to the book. Don't worry about the answers, just try to remember how to show someone else the trick.

I will tell you that you will get a lot more out of it if you write down what I'm saying on a piece of paper or a white board.

You'll get even more out of it if you share all the tips and tricks in this book to help someone else get better at multiplication.

Good luck.

Tony Guerra

tonythepharmacist@gmail.com

MATH AND THE DISHWASHING MACHINE

As part of their chores, my triplet daughters empty and put away the plates, cups and so on from the dishwasher. Each triplet will do their part of the dishwasher and ONLY their part of the dishwasher.

One puts away the dishes on the bottom of the dishwasher, the easiest job, one will put away the silverware, the second easiest job and the last will put away the top of the dishwasher the hardest job since that's where all the stuff goes that no one knows where it belongs AND has to put the dishes that were in the sink into the dishwasher, so that job takes a lot longer.

Since the dishwasher runs at night, it's ready to be emptied first thing in the morning. That means that the person doing the easiest, middle, and hardest parts of the dishwasher are generally the first, second, and last to wake up and that really doesn't change much. What does this have to do with the times table?

If the my triplets split up the times table like they split up the dishwasher and other chores, triplet A will only work with numbers 1, 2, 3, and 4. Triplet B will only take numbers 5, 6, 7, and eight, and triple C will take numbers 9, 10, 11, and 12.

So when I'm referring to the triplet A, triplet B or triplet C, I'm really referring to the numbers that they're responsible for. Again it's

Triplet A – Early riser, easiest: 1, 2, 3, 4

Triplet B – Middle, a little harder: 5, 6, 7, 8

Triplet C – Late riser, hardest, 9, 10, 11, 12

But that is a lot of exhausting brain work to do when the kid next to me just says, "Six times nine equals fifty-four" (6 x 9 = 54) without all those brain gymnastics. It was also discouraging because I was slow to finish my math tests, and my classmates were fast. The more I compared myself to other people, and I know they say you shouldn't do that, but I did it anyway, the more I got discouraged until I said, "I'm not good at math."

Eventually, I figured out a way to learn the times tables once I became a tutor and then a teacher and then a professor, but I wish I knew this trick back then, so I want to share it with you because I've seen how painful being behind in math can be.

We have a saying in teaching that goes, "I never really learned it until I taught it" and I believe that's true, you aren't an expert at something until you try to teach it to someone else.

Let me tell you a story.

One day, I forgot my notes when I was supposed to teach a class. I'm kind of busy with three kids, a dog, an adopted cat, and two foster cats, so it happens. But I was able to teach the class anyway because I remembered most everything. How?

Because I taught it many times, my brain remembered it. That's when the lightbulb went off in my head. For the multiplication table to stick to my brain like chewing gum sticks to your hair, I had to learn to teach it from memory. I had the wrong goal.

My goal shouldn't be to *remember* the times table, my goal was creating a *memorable* times table that would stick and that's what this book does. I stopped feeling alone in a crowded classroom as now I was helping other people. I got even better explaining why sometimes the learning was slow and sometimes the learning was fast and could help others feel better.

When you start learning the times tables, like with ones and twos, you learn fast. Then, you might do skip counting for

We could do these multiplication sets one at a time or two at a time, but I found that is too slow. You don't feel like you are making progress. Let me tell you a quick story.

When I did a four-building climb challenge where I went up 84 flights of stairs in four skyscrapers, I would see some people going up one step at a time. Other people would skip a step and go a little faster Others would use one arm on the handrail and go even faster. And others, like me, skipped one step and used both hands on the handrails to get up the fastest. Using those two steps and two hands was just the right speed.

In the same way, by going through the times table four at a time, you'll find it's just the right amount. That's the whole reason you memorize the times table in the first place. It's a lot faster to answer questions if you have the product in your memory than to try to even use a calculator for single digit multiplication. Before we start, let's have a quick chat as to why multiplication is hard in the first place.

WHY IS MULTIPLICATION HARD?

I never learned my times tables all the way when I was supposed to in elementary school. Something happened, maybe I got sick, but I just remember that I only got to about the six times table and I didn't really get far in that. I only got to about six times six equals thirty-six ($6 \times 6 = 36$).

So, when someone asked me what six times eight equals, in my head I had to say, "Six times six equals thirty-six plus six more equals forty-two plus six again equals forty-eight." ($6 \times 6 = 36 + 6 = 42, + 6 = 48$)

If someone asked me what six times nine equals, it would be a little easier because I would say six times ten is sixty, minus six equals fifty-four. ($6 \times 10 = 60 - 6 = 54$)

Introduction – 0s, All Zeros

First let's learn the setup. We're going to always put the number we are multiplying by first, in this case it's zero. The **Zero Property** states that the product of any number and 0 is 0. Triplet A with numbers 1 to 4 would say that:

0 x 1 = 0

0 x 2 = 0

0 x 3 = 0

0 x 4 = 0

Triplet B responsible for numbers 5 through 8 would say:

0 x 5 = 0

0 x 6 = 0

0 x 7 = 0

0 x 8 = 0

Finally, Triplet C would round out with 9 through 12.

0 x 9 = 0

0 x 10 = 0

0 x 11 = 0

0 x 12 = 0

Let's move on to the ones.

Chapter 1 – 1s, Same Number

The **Identity Property** states that the product of any number and 1 is that number, so Triplet A with 1 to 4 would say:

1 x 1 = 1

1 x 2 = 2

1 x 3 = 3

1 x 4 = 4

Triplet B responsible for numbers 5 through 8 would say:

1 x 5 = 5

1 x 6 = 6

1 x 7 = 7

1 x 8 = 8

Finally, Triplet C would round out with 9 through 12.

1 x 9 = 9

1 x 10 = 10

1 x 11 = 11

1 x 12 = 12

Again, we're just focusing on making sure we memorize the setup.

The chapter number is always the first number, the triplet's assigned four numbers are always second. What technique will the triplets use to multiply twos?

CHAPTER 2 – 2S, ADD ITSELF

The twos rule is that any number multiplied by two can be added to itself to get the same answer. For example, 2 x 3 = 6. But 3 + 3 = 6, also. Instead of writing that 2 x 3 = 6, which is not memorable, we write 2 x 3 = 3 + 3 = 6. So, Triplet A with numbers 1 to 4 would say that:

2 x 1 = 1 + 1 = 2

2 x 2 = 2 + 2 = 4

2 x 3 = 3 + 3 = 6

2 x 4 = 4 + 4 = 8

Triplet B responsible for numbers 5 through 8 would say:

2 x 5 = 5 + 5 = 10

2 x 6 = 6 + 6 = 12

2 x 7 = 7 + 7 = 14

2 x 8 = 8 + 8 = 16

Finally, Triplet C would round out with 9 through 12.

2 x 9 = 9 + 9 = 18

2 x 10 = 10 + 10 = 20

2 x 11 = 11 + 11 = 22

2 x 12 = 12 + 12 = 24

What technique will the triplets use in Chapter 3 to figure out how to multiply the threes?

Chapter 3 – 3s, Skip Counting and More

This chapter is one where you'll use your hands. Before we start with the 3s, we should identify a number from one to ten associated with each finger. In school, we are often taught to put our hands down on the table and trace our hands on a piece of paper and write the numbers one through ten above our fingers. While I am writing the words out as o-n-e, and t-w-o here, I recommend using the digits 1 and 2 and so on.

1. Above our left pinkie or little finger, we'll put a one.
2. Above our left ring finger, we'll put a two.
3. Above our left middle or long finger, we'll put a three.
4. Above our left index or pointer finger, we'll put a four.
5. Above our left thumb, we'll put a five.

Now we'll do the right hand.

6. Above our right thumb, we'll put a six.
7. Above our right index finger, we'll put a seven.
8. Above our right middle finger, we'll put an eight.
9. Above our right ring finger, we'll put a nine.
10. Above our right pinkie, we'll put a ten.

Don't worry, we have a strategy for eleven and twelve, but we'll wait for Triplet C to help us out with that. Now, the temptation is to quickly do one through ten.

But our short-term memory doesn't like having to deal with more than three to seven things at a time.

So, in this chapter, we'll do the same as we've done before letting Triplet A handle one through four, Triplet B five through eight and Triplet C handle nine through twelve.

TRIPLET A – 3S, SKIP COUNTING – 1 TO 4

Skip counting is literally skipping numbers to arrive at the next number. Instead of counting by one from one to ten as 1, 2, 3, 4, 5, 6, 7, 8, 9, 10 we're going to skip count from three to thirty as 3, 6, 9, 12, 15, 18, 21, 24, 27, and 30.

Because we're breaking it up by triplet responsibilities, we're only going to focus on multiplying three times one, two, three and four to begin with. Note, there is only one syllable in three, six, nine, and twelve (3, 6, 9, 12) so that makes it easier and punchier to remember.

As we count off the numbers, we can imagine the numbers above our fingers. However, instead of writing that three times two equals six (3 x 2 = 6) we will write out the method so three times two equals three comma six (3 x 2 = 3, 6) knowing that the last number we get to after Skip counting is our answer.

So, Triplet A would do her job by memorizing:

3 x 1 = 3

3 x 2 = 3, 6

3 x 3 = 3, 6, 9. (You can picture a nine square tic-tac-toe board)

3 x 4 = 3, 6, 9, 12

TRIPLET B –3S, SKIP COUNTING – 5 TO 8

Let's first see what it sounds like when we skip count the 3s in the same way Triplet A did for numbers one through four.

I think you'll find it's way too much work and that we'll want to look for some kind of shortcut to make it easier on us.

Following the same pattern as Triplet A used, Triplet B would say:

3 x 5 = 3, 6, 9, 12, 15

3 x 6 = 3, 6, 9, 12, 15, 18

3 x 7 = 3, 6, 9, 12, 15, 18, 21

3 x 8 = 3, 6, 9, 12, 15, 18, 21, 24

But Triplet B's not interested in counting out that many numbers. So, she looks at the answers to three times five equals fifteen (3 x 5 = 15) and three times six equals eighteen (3 x 6 = 18) and sees that unlike the answers to three times one through four: three, six, nine, twelve (3, 6, 9, 12) which each have one syllable that fifteen and eighteen (15 and 18) have two syllables.

So, Triplet B will look at her left thumb and over-emphasize the accent on the fif-teen and her right thumb and over-emphasize the eight-teen to help her memorize them. In that way, she can start her skip counting at fifteen, eighteen.

To add more contrast twenty-one and twenty-four each have three syllables. So, her responsibility starts with her left and right thumbs each with two syllables as:

3 x 5 = 15

3 x 6 = 15, 18

3 x 7 = 15, 18, 21

3 x 8 = 15, 18, 21, 24

We are using an anchor calculation that is easy to remember. Memorizing 3 x 5 = 15 starts us off quickly. But what will Triplet C do? She wants nothing to do with skip counting so many numbers. She'll use a completely different technique that's even faster and will help with other numbers.

TRIPLET C – 3S, SUBTRACTION, ADDING ZERO, TWINS, TIMES 10, TIMES 2 – 9 TO 12

Let's first see what it sounds like when we skip count the 3s in the same way Triplet B did for numbers five through eight. I think you'll find, again, it's way too much work and that we'll want to look for some kind of shortcut to make it easier on us.

3 x 9 = 15, 18, 21, 24, 27

3 x 10 = 15, 18, 21, 24, 27, 30

3 x 11 = 15, 18, 21, 24, 27, 30, 33

3 x 12 = 15, 18, 21, 24, 27, 30, 33, 36

How is Triplet C ever going to get back to her iPad and playing Roblox if she's counting all these numbers?

Triplet C realizes that even easier than multiplying three times five to get fifteen is multiplying a number times ten. Any number times ten equals that number with a zero after it. So, three times ten equals three zero or thirty (3 x 10 = 3 0 or 30).

No one said that your **anchor calculation**, the easiest calculation you go back to if you forget the answer, had to be the first multiplication set you have.

Sure, Triplet A's easiest calculation, her anchor calculation, is three times one equals three (3 x 1 = 3).

Triplet B's easiest calculation, her anchor calculation, is three times five equals fifteen (3 x 5 = 15).

But Triplet C's easiest calculation isn't three times nine equals twenty-seven (3 x 9 = 27), it's three times ten equals thirty (3 x 10 = 30).

But Triplet C has more tricks up her sleeve than just that, she's going to use a special technique with three times nine, three times eleven, and three times twelve also.

For three times nine equals twenty-seven (3 x 9 = 27) Triplet C realizes that subtraction is easier than multiplication. So, of her four responsibilities, it would be easiest to multiply three times ten to make thirty then subtract three to get the answer to three times nine as twenty-seven (3 x 10 = 30 - 3 = 27).

For three times eleven, Triplet C knows that any number between one and nine multiplied by eleven is that number twice, so you don't get triplets, but twins. Three time eleven equals three, three or thirty-three (3 x 11 = 3 3 = 33) or twin threes.

For three times twelve Triplet C uses a clever addition of two techniques. When you think about it, a twelve is really a ten and a two. So, if you multiply three times ten equals thirty (3 x 10 = 30) and three times two equals six (3 x 2 = 6) and add those products together, you get thirty-six. While sometimes the triplets do things the same, sometimes they veer off on their own. Here's the final strategies from Triplet C.

3 x 9 = 30 - 3 = 27

3 x 10 = 3 0 = 30

3 x 11 = 3 3 = 33

3 x 12 = (3 x 10) + (3 x 2) = 30 + 6 = 36

Triplet C's 9, 10, 11, and 12 were supposed to be the hardest, but with these shortcuts, they can also be some of the easiest. We've gone over a lot in this chapter, maybe we should review. Triplet A used her left hand to skip count:

3 x 1 = 3

3 x 2 = 3, 6

3 x 3 = 3, 6, 9

3 x 4 = 3, 6, 9, 12

Triplet B started with her left thumb at fif-teen and right thumb at eight-teen to say:

3 x 5 = 15

3 x 6 = 15, 18

3 x 7 = 15, 18, 21

3 x 8 = 15, 18, 21, 24

Finally, Triplet C went in a different direction with some clever shortcuts to get:

3 x 9 = 30 − 3 = 27

3 x 10 = 3 0 = 30

3 x 11 = 3 3 = 33

3 x 12 = (3 x 10) + (3 x 2) = 30 + 6 = 36

Will we be able to use these same techniques with the 4s?

Chapter 4 – 4s, Skip Counting and More

We can also do skip counting with the four times tables, but right about now you might be getting a little bit bored. What I like to do is think about something that I care about – horses.

We traditionally measure the height of horses in Great Britain and the United States in hands. A single hand as a measure equals four inches. You measure from the ground to where the neck and back meet at the top of the shoulders, also known as the withers. To picture the height of the world's shortest horse you need to be able to multiply four inches times four hands to equal sixteen inches. The horse is a tiny bit taller than that.

The tallest horse ever measured was a little over 21 hands. If you multiply 21 times 4, you get 84 inches or seven feet. That is also easy to picture. The important thing is that you start caring about the math because you care about what it can tell you.

If we want to, we can put your hands down on the table and trace our hands on a piece of paper and write the numbers one through ten (1 through 10) above our fingers as we've done before.

While I am writing the entire words one, o-n-e, and two, t-w-o, and so forth here in the printed version, I recommend using the digits 1, 2, 3 and so forth as you did with the threes.

1. Above our left pinkie or little finger, we'll put a one.
2. Above our left ring finger, we'll put a two.
3. Above our left middle or long finger, we'll put a three.
4. Above our left index or pointer finger, we'll put a four.
5. Above our left thumb, we'll put a five.

Now we'll do the right hand.

You can do this while you are petting your dog or cat or other pet, and you are listening to stories about our new Vizsla Beau, B-e-a-u and his adventures with math. Is Beau real?

Yes, he is, he's lying on the couch next to me waiting for me to run with him. He's comfortable running up to 14 miles with me on the bike trail.

In this book, I will break the times table up into manageable bites of four calculations at a time with the help of some of my family members. I picked that number four, f-o-u-r, for, f-o-r a reason, it's easy on your brain and keeps you relaxed.

When you remember a phone number, you don't try to remember all ten numbers, you chunk it to three numbers, the area code, then the first three numbers and the last four numbers. So, area code five-one-five, then one, two, three, then, four, five, six, seven would be chunking the phone number 515-123-4567.

When you look at a credit card, you see four numbers, then a space, then four numbers, then a space, then four numbers, a final space, and a last four numbers.

Why don't credit card companies just put the sixteen numbers together? It's hard to distinguish the numbers and hard to remember. It hurts your brain if you do it that way.

In the same way, the traditional times table or multiplication table from 1 to 12 is, for lack of a better word, too squished. It hurts your brain and is hard to remember.

We don't want to try to remember 12 things at once, let's take a manageable number like four, which is right in the middle of what your brain will allow for short term memory, somewhere between three and seven items.

AUTHOR'S NOTE

WHAT IS A RELAXED APPROACH?

What would make learning multiplication a lot more relaxing?

A dog.

When we got our first Vizsla Talladega, or Tally, it was my first dog, and I didn't know how she would do with my young triplet daughters. What I found was unexpected. When one of the triplets had trouble, they would cuddle up next to Talladega and she ended up being a comfort dog.

Normally, comfort dogs are used to comfort someone who has experienced a tragic or traumatic event in the past. Well, I thought, you know, to many people, math is going to be a traumatic event in the making if it hasn't been already. Maybe, I could write a book that would comfort someone as they were going through what to them, is a traumatic event – math.

I'm Tony, I teach chemistry and pharmacology in college, but there's a lot of math in those subjects that make my students very uncomfortable. I realized that I needed to take a step back and help my students get over a mountain many of them never actually got over – thinking they were bad at math.

What I've done in this book is teach the times tables in a way that is much more relaxing. You don't need to write anything down. You don't need to do countless worksheets. This is an audiobook script meant to be read or listened to straight through. Just read or listen to how I break up the problem of having to remember the 144 problems on a 1 to 12 times table.

Table of Contents

Chapter 11 – 11s, Twins and Split, Add, Insert (Spain Mnemonic) 46
    Triplet A – 11s, Twins - 1 to 4 46
    Triplet B – 11s, Twins – 5 to 8 46
    Triplet C – Twins, Add a Zero, Split, Add, and Insert (SPAIN Mnemonic) 47
Chapter 12 – 12s, Times 10, Times 2 and Add Them 50
    Triplet A – 12s, Times 10, times 2 and Add Them - 1 to 4 50
    Triplet B– 12s, Times 10, times 2 and Add Them - 5 to 8 50
    Triplet C– 12s, Times 10, times 2 and Add Them - 9 to 12 51
Chapter 13 – Multiplying Doubles with Rhymes 53
    Triplet A – Doubles Rhymes - 1 to 4 53
    Triplet B – Doubles Rhymes - 5 to 8 54
    Triplet C – Doubles Rhymes - 9 to 12 55
    Shortening the Rhymes 56
Afterword 58

6. Above our right thumb, we'll put a six.
7. Above our right index finger, we'll put a seven.
8. Above our right middle finger, we'll put an eight.
9. Above our right ring finger, we'll put a nine.
10. Above our right pinkie or little finger, we'll put a ten.

Now we're ready to skip count by fours.

So, instead of counting by one as one, two, three, four, five, six, seven, eight, nine, and ten we'll count by fours and pay attention to the number of syllables.

There is only one syllable in four, eight, and twelve (4, 8, and 12). There are two syllables in sixteen and twenty (16 and 20). There are three syllables in twenty-four, twenty-eight, thirty-two, and thirty-six (24, 28, 32, and 36). Finally, there are two syllables in forty (40).

Now, Triplet C will probably use a different method for four times nine and four times ten that's easier, but when I didn't mention all the numbers and all the fingers, I couldn't help but feel like I was missing some fingers which was terrifying.

TRIPLET A – SKIP COUNTING 4S – 1 TO 4

So, the fours cadence goes like this, moving your fingers as you count: four, eight, twelve (4, 8, 12) and sixteen overemphasizing the six-teen with your left pointer finger.

So, Triplet A would do her job by memorizing:

4 x 1 = 4

4 x 2 = 4, 8

4 x 3 = 4, 8, 12

4 x 4 = 4, 8, 12, 16

Later in the book we'll talk about some ways to memorize the doubles like 2 x 2, 3 x 3, and 4 x 4 with rhymes to make this even faster.

TRIPLET B – SKIP COUNTING 4S – 5 TO 8

We saw with the threes that it doesn't make sense to skip count all the way from one, rather, we work to memorize the numbers on the thumbs. The nice thing about multiplying 5 times anything is that you know the last number will be either a 5 or a 0, as 1 5, 15 or 2 0, 20.

So, we'll memorize our left thumb as 4 x 5 = 20. The right thumb is then 4 x 6 = 20, 24 where eventually it will stay in your brain that 4 x 6 = 24. Then we add another 4 to make 4 x 7 = 20, 24, 28 and finally 4 x 8 = 20, 24, 28, 32.

So, her responsibility starts with her left and right thumbs as:

4 x 5 = 20

4 x 6 = 20, 24

4 x 7 = 20, 24, 28

4 x 8 = 20, 24, 28, 32

The most important thing is that we use the anchor of 4 x 5 = 20 to start us off and this way we don't have to skip count all the way from 4 x 1 = 4. But what will Triplet C do this time? She still wants nothing to do with skip counting so many numbers. She'll use her same technique from the threes.

Triplet C – 4s, Subtraction, Adding Zero, Twins, Times 10, times 2 – 9 to 12

For four times nine equals thirty-six (4 x 9 = 36) Triplet C realizes that again subtraction is easier than multiplication. So, of her four responsibilities, it would be easiest to just multiply four times ten to make forty (4 x 10 = 40 – 4 = 36) then subtract four to get the answer to four times nine which is thirty-six.

For four times eleven, Triplet C pulls out her twins again knowing that four times eleven equals four, four or forty-four (4 x 11 = 4 4 = 44) or twin fours.

For the number four times twelve Triplet C again uses the addition of two techniques. We talked about how the twelve is really a ten and a two. So, if you multiply four times ten equals forty (4 x 10 = 40) and four times two equals eight (4 x 2 = 8) and add those products together, you get forty-eight. Here's the final strategies from Triplet C as she conquers the fours.

4 x 9 = 40 – 4 = 36

4 x 10 = 4 0 = 40

4 x 11 = 4 4 = 44

4 x 12 = (4 x 10) + (4 x 2) = 40 + 8 = 48

We've gone over a lot in this chapter, maybe we should review. Triplet A used her left hand to skip count:

4 x 1 = 4

4 x 2 = 4, 8

4 x 3 = 4, 8, 12

4 x 4 = 4, 8, 12, 16

Triplet B started with her left thumb at twen-ty and right thumb at twen-ty FOUR to say:

4 x 5 = 20

4 x 6 = 20, 24

4 x 7 = 20, 24, 28

4 x 8 = 20, 24, 28, 32

Finally, Triplet C went in a different direction with some clever shortcuts to get:

4 x 9 = 40 – 4 = 36

4 x 10 = 4 0 = 40

4 x 11 = 4 4 = 44

4 x 12 = (4 x 10) + (4 x 2) = 40 + 8 = 48

Will the fives be a little bit of a break because the products all end in 5 or 0?

CHAPTER 5 – 5S, SKIP COUNTING AND MORE

Like with the numbers three and four, you can skip count with fives. What's special about fives, however, is that because multiplying by an odd number results in a number that ends in five and an even number a number that ends in zero, most find it easy to just say five, ten, fifteen, twenty, twenty-five, thirty and so on (5, 10, 15, 20, 25, 30).

We can do skip counting as with the four times tables, but we'll also learn another strategy as well after we've gone through all of five times one (5 x 1) through five times ten (5 x 10). Just like with the three times tables and four times tables, we'll put our hands down on the table and trace our hands on a piece of paper and write the numbers one through ten (1 through 10) above our fingers and we'll let Triplet C deal with 11 and 12.

1. Above our left pinkie or little finger, we'll put a one.
2. Above our left ring finger, we'll put a two.
3. Above our left middle or long finger, we'll put a three.
4. Above our left index or pointer finger, we'll put a four.
5. Above our left thumb, we'll put a five.

Now we'll do the right hand.

1. Above our right thumb, we'll put a six.
2. Above our right index finger, we'll put a seven.
3. Above our right middle finger, we'll put an eight.
4. Above our right ring finger, we'll put a nine.
5. Above our right pinkie or little finger, we'll put a ten.

Now we're ready to skip count by 5s. So instead of counting by 1 from 1 to 10, we would count from 5 to 50, but let's see how the triplets handle these.

TRIPLET A – SKIP COUNTING 5S – 1 TO 4

There is only one syllable in five and ten (5 and 10). There are two syllables in fifteen and twenty (15 and 20).

So, the 5s cadence for the first four numbers goes like this, moving your fingers as you count: 5, 10 and then 15, 20.

So, Triplet A would do her job by memorizing:

5 x 1 = 5

5 x 2 = 5, 10

5 x 3 = 5, 10, 15

5 x 4 = 5, 10, 15, 20

However, what we find is that students tend to just put these to memory very quickly without the skip counting because the last digit, the 5 or 0, is so consistent.

TRIPLET B – SKIP COUNTING 5S – 5 TO 8

So, we'll memorize our left thumb as 5 x 5 = 25, which has three syllables. The right thumb is then 5 x 6 = 30, two syllables.

Then the syllables alternate again so it's 5 x 7 is 35, which has three syllables and 5 x 8 = 40 which has two syllables.

So, her responsibility starts with her left and right thumbs as:

5 x 5 = 25

5 x 6 = 25, 30

5 x 7 = 25, 30, 35

5 x 8 = 25, 30, 35, 40

The most important thing is that we use the anchor of 5 x 5 = 25. Later we'll use a rhyme to remember the doubles like 5 x 5, but you can use it now if you like that 5 x 5 is too many dives, 5 x 5 is twenty-five.

TRIPLET C – 5S, SUBTRACTION, ADDING ZERO, TWINS, TIMES 10, TIMES 2 – 9 TO 12

For five times nine equals forty-five (5 x 9 = 45) Triplet C realizes that again subtraction is easier than multiplication. So, of her number five responsibilities, it would be easiest to just multiply five times ten to make fifty then subtract five (5 x 10 = 50 – 5 = 45) to get the answer to five times nine which is forty-five.

For five times eleven, Triplet C pulls out her twins again knowing that five times eleven equals five, five or fifty-five (5 x 11 = 5 5 = 55) or twin fives. For five times twelve Triplet C again uses the addition of two techniques. We talked about how the twelve is really a ten and a two. So, if you multiply five times ten equals fifty (5 x 10 = 50) and five times two equals ten (5 x 2 = 10) and add those products together, you get 50 + 10 = 60. Here's the final strategies from Triplet C as she conquers the fives.

5 x 9 = 50 – 5 = 45

5 x 10 = 5 0 = 50

5 x 11 = 5 5 = 55

5 x 12 = (5 x 10) + (5 x 2) = 50 + 10 = 60

We've gone over a lot in this chapter, maybe we should review.

Triplet A used her left hand to skip count:

5 x 1 = 5

$5 \times 2 = 5, 10$

$5 \times 3 = 5, 10, 15$

$5 \times 4 = 5, 10, 15, 20$

Triplet B started with her left thumb at twen-ty FIVE and right thumb at thir-ty to say:

$5 \times 5 = 25$

$5 \times 6 = 25, 30$

$5 \times 7 = 25, 30, 35$

$5 \times 8 = 25, 30, 35, 40$

Finally, Triplet C went in a different direction with her clever shortcuts to get:

$5 \times 9 = 50 - 5 = 45$

$5 \times 10 = 5\ 0 = 50$

$5 \times 11 = 5\ 5 = 55$

$5 \times 12 = (5 \times 10) + (5 \times 2) = 50 + 10 = 60$

While multiplying sixes is generally harder than multiplying fives, we'll see that Triplet B has a clever way to remember the toughest multiplications with six.

Chapter 6 – 6s, Double 3s and Rhymes

Multiplying by threes and twos is a lot easier than multiplying by six. If you think about it, we learned in Chapter 2 that 2 x 3 = 6 and in Chapter 3 that 3 x 2 = 6, so we can get the exact same answer if we multiply a number by three and then by two as we would multiplying by six. For example, 6 x 7 = 3 x 7 = 21 (pause) x 2 = 42. And 6 x 7 does equal 42.

Triplet A – 6s, Double 3s and Rhymes – 1 to 4

So, Triplet A could do her job by memorizing that:

6 x 1 = 3 x 1 = 3 (pause) x 2 = 6

6 x 2 = 3 x 2 = 6 (pause) x 2 = 12

6 x 3 = 3 x 3 = 9 (pause) x 2 = 18

6 x 4 = 3 x 4 = 12 (pause) x 2 = 24

You've already learned 2 x 3, 2 x 6, 2 x 9, and 2 x 12 so flipping them around to make 3 x 2, 6 x 2, 9 x 2, and 12 x 2 shouldn't be too bad. But we challenged Triplet A to come up with some rhymes that include her copper coated Vizsla pointer dog Beau. Triplet A came up with:

6 x 1, Your face Beau *licks*, so 6 x 1 = 6.

6 x 2, Beau jumps on a *shelf*, so 6 x 2 = 12.

6 x 3, Beau's eyes are *green*, so 6 x 3 = 18.

6 x 4, Beau runs *for* the *door*, so 6 x 4 = 24.

Triplet B – 6s, Double 3s or Rhymes – 5 to 8

So, Triplet B could do something very similar to Triplet A with doubling the answer after multiplying by three but we to run into some bigger numbers:

6 x 5 = 3 x 5 = 15 (pause) x 2 = 30

6 x 6 = 3 x 6 = 18 (pause) x 2 = 36

6 x 3 = 3 x 7 = 21 (pause) x 2 = 42

6 x 4 = 3 x 8 = 24 (pause) x 2 = 48

But now we are multiplying numbers that are higher than 12 and it might be easier to just use Triplet B's rhymes.

6 x 5, Beau got so *dirty*, so 6 x 5 = 30.

6 x 6, Beau picks up *six* sticks, so 6 x 6 = 36.

6 x 7, This week Beau had seven *poos*, so 6 x 7 = 42.

6 x 8, Beau's got a play *date*, so 6 x 8 = 48.

Yeah, those rhymes seem a lot easier. By the way, dogs have 42 teeth, but that didn't rhyme with 6 times 7.

Triplet C – 6s, Subtraction, Adding Zero, Twins, Times 10, times 2 – 9 to 12

Let's see what Triplet C would look like if she followed the double three method.

6 x 9 = 3 x 9 = 27 (pause) x 2 = 54

6 x 10 = 3 x 10 = 30 (pause) x 2 = 60

6 x 11 = 3 x 11 = 33 (pause) x 2 = <u>66</u>

6 x 12 = 3 x 12 = 36 (pause) x 2 = 72

Again, we run into the problem of multiplying big numbers. Let's see what happens when Triplet C uses her usual method of subtracting a six, adding a 0, twins with 11, and times 10 times 2 with 12.

6 x 9 = 60 – 6 = 54

6 x 10 = 6 0 = 60

6 x 11 = 6 6 = 66

6 x 12 = (6 x 10) + (6 x 2) = 60 + 12 = 72

Those aren't that bad, but the one thing triplets hate is having to do something their sister doesn't do, so Triplet C needs to come up with some rhymes too.

6 x 9, Beau's tired and *snores*, so 6 x 9 = 54

6 x 10, Gross, again Beau *licked me*, so 6 x 10 = 60

6 x 11, Beau can do lots of *tricks*, so 6 x 11 = 66

6 x 12, Beau *chews two shoes*, so 6 x 12 = 72

Now, instead of reviewing the double threes method, let's think of a pet that you own, and you can put the name of the pet where Beau's name was. For example, I'll just change the name to Arty for our adopted tabby cat Aristotle's nickname. Do realize Arty the cat wouldn't do all the things a dog would do. It makes it more memorable when you picture a cat picking up six sticks 36, play date 48, doing tricks 66, or chewing two shoes 72 as those are things our Arty would never do. Here we go.

6 x 1, Your face Arty *licks*, so 6 x 1 = 6.

6 x 2, Arty jumps on a *shelf*, so 6 x 2 = 12.

6 x 3, Arty's eyes are *green*, so 6 x 3 = 18.

6 x 4, Arty runs *for* the *door*, so 6 x 4 = 24.

(pause)

6 x 5, Arty got so *dirty*, so 6 x 5 = 30.

6 x 6, Arty picks up *six* sticks, so 6 x 6 = 36.

6 x 7, Arty had seven *poos*, so 6 x 7 = 42.

6 x 8, Arty got a play *date*, so 6 x 8 = 48.

(pause)

6 x 9, Arty's tired and *snores*, so 6 x 9 = 54

6 x 10, Gross, again Arty *licked me*, so 6 x 10 = 60

6 x 11, Arty does lots of *tricks*, so 6 x 11 = 66

6 x 12, Arty *chews two shoes*, so 6 x 12 = 72

Now, we're kind of at the point where the Commutative Property of Multiplication really helps us out where 6 x 7 is the same as 7 x 6 or more technically, changing the order of factors does not change their product. We've already learned:

1 x 7 = 7.

2 x 7 = 7 + 7 = 14.

3 x 7 = 15, 18, 21, starting with our anchor 3 x 5 = 15.

4 x 7 = 20, 24, 28 starting with our anchor 4 x 5 = 20.

5 x 7 = 25, 30, 35, starting with our anchor 5 x 5 = 25.

6 x 7, Beau has seven poos, 6 x 7 = 42.

So, before we even start the next chapter, we're halfway there.

Chapter 7 – 7s, Skip Counting, Rhymes, and More.

Even though we have a technique for everything from 1 to 6 times 7, sometimes it's easier to use another method when things get flipped around.

Triplet A – 7s, Skip Counting – 1 to 4

Triplet A decided that she liked skip counting for 7 x 1 through 7 x 4 and I agree, with those smaller numbers, it makes it easier to memorize.

7 x 1 = 7

7 x 2 = 7, 14

7 x 3 = 7, 14, 21

7 x 4 = 7, 14, 21, 28

Triplet B – 7s, Anchor and Rhymes – 5 to 8

Triplet B, however, finds that she remembers 5 x 7 = 35 and 7 x 5 = 35 easily, so she just puts that one down as:

7 x 5 = 5 x 7 = 35

However, she uses rhymes for 7 x 6, 7 x 7 and 7 x 8.

7 x 6 = 6 x 7 where in a week Beau has 7 poos, 42.

7 x 7, however, gets a new rhyme. If you look at two sevens, they are made up of two lines each, one horizontal and one diagonal. So, our rhyme goes 7 x 7, have *four* straight *lines*, so 7 x 7 is *49*.

7 x 8 = 56 is a little bit tougher. You could do a good job remembering that 7 x 7 = 49 and add another 7 to get 56, but it seems like a rhyme is best here.

7 x 8, Beau has a good diet and doesn't lick bricks, 7 x 8 = 56.

Some dogs will lick bricks if their diet lacks certain nutrients. Dogs, with their excellent sense of smell, are sensing that the brick may have the minerals they need. Also, sometimes dogs that have cancer, anemia or others stomach issues will like licking bricks. Hopefully now you'll remember Beau has a good diet and doesn't lick bricks, 7 x 8 = 56.

So, to summarize,

7 x 5 = 5 x 7 = 35

7 x 6 = 6 x 7 where in a week Beau has 7 poos, 42.

7 x 7, have *four* straight *lines*, so 7 x 7 is *49*.

7 x 8, Beau eats well and doesn't eat bricks, 7 x 8 = 56.

TRIPLET C – 7S, SUBTRACTION, ADDING ZERO, TWINS, TIMES 10, TIMES 2 – 9 TO 12

Triplet C wants nothing to do with that complicated brick talk and rhymes and goes back to her tried and true methods.

7 x 9 = 70 – 7 = 63

7 x 10 = 7 0 = 70

7 x 11 = 7 7 = 77

7 x 12 = (7 x 10) + (7 x 2) = 70 + 14 = 84

Let's summarize this very intense chapter that combines skip counting, rhymes, and more.

7 x 1 = 7

7 x 2 = 7, 14

7 x 3 = 7, 14, 21

7 x 4 = 7, 14, 21, 28

(pause)

7 x 5 = 5 x 7 = 35

7 x 6 = 6 x 7 where in a week Beau has 7 poos, 42.

7 x 7, have *four* straight *lines*, so 7 x 7 is *49*.

7 x 8, Beau eats well and doesn't lick bricks, 7 x 8 = 56.

(pause)

7 x 9 = 70 – 7 = 63

7 x 10 = 7 0 = 70

7 x 11 = 7 7 = 77

7 x 12 = (7 x 10) + (7 x 2) = 70 + 14 = 84

Chapter 8 – 8s, Double 4s.

When we tried the double threes with the sixes, it seemed to be quite hard because you can get even and odd numbers.

When you multiply an odd number, 3, times an odd number, 7, or 3 x 7 you get 21, an odd number.

But if you multiply an odd number, 3, times an even number, 8, you get 24, an even number.

However, when you multiply an even number like 8 times an odd number, 3, or an even number, 4, you'll get 24 and 32, both even numbers. So, every answer that is multiplied by eight will end in 2, 4, 6, 8, or 0. Let's see how this works and if we can detect an even more helpful pattern.

Triplet A – 8s, Skip Counting and Double 4s – 1 to 4

Triplet A starts off with skip counting the eights and is looking for a pattern.

8 x 1 = 8

8 x 2 = 8, 16

8 x 3 = 8, 16, 24

8 x 4 = 8, 16, 24, 32

The second digit in the ones place are counting down by twos as 8, 6, 4, 2.

So, when we use the times 4, times 2 method we see how this works in the same way.

8 x 1 = 4 x 1 = 4 (pause) x 2 = 8, ends with 8

8 x 2 = 4 x 2 = 8 (pause) x 2 = 16, ends with 6

8 x 3 = 4 x 3 = 12 (pause) x 2 = 24, ends with 4

8 x 4 = 4 x 4 = 16 (pause) x 2 = 32, ends with 2

The ones place digits go backward as 8, 6, 4, 2.

TRIPLET B – 8S, SKIP COUNTING AND DOUBLE 4S – 5 TO 8

Triplet B, finds that she has no problem remembering that:

8 x 5 = 40 because she knows her five times table well.

However, if we want to stay with the Double 4s method, she can say:

8 x 5 = 4 x 5 = 20 (pause) x 2 = 40.

8 x 6 = 4 x 6 = 24 (pause) x 2 = 48.

8 x 7 = 4 x 7 = 28 (pause) x 2 = 56

8 x 8 = 4 x 8 = 36 (pause) x 2 = 72

But if those multiplications of double digits are getting hard, we can go back to adding rhymes again and use that Commutative Property of Multiplication to help us.

8 x 6 = 6 x 8, Beau got a play date, so 6 x 8 = 48.

8 x 7 = 7 x 8, Beau eats well and doesn't lick bricks, 7 x 8 = 56.

8 x 8 requires a new rhyme as we haven't done one with these two numbers together before. In this rhyme there is "ate, a-t-e" like Beau ate a treat and "eight, e-i-g-h-t" as in Beau was one of

eight puppies. These are homophones, which means they have different meanings but sound the same.

8 x 8, Beau ate and ate, got sick on the floor, so 8 x 8 is sixty-four.

8 x 5 = 4 x 5 = 20 (pause) x 2 = 40

8 x 6 = 6 x 8, Beau got a play date, so 6 x 8 = 48.

8 x 7 = 7 x 8, Beau eats well and doesn't lick bricks, 7 x 8 = 56.

8 x 8, Beau ate and ate, got sick on the floor, 8 x 8 is sixty-four.

Triplet C – 8s, Subtraction, Adding Zero, Twins, Times 10, Times 2 – 9 to 12

Triplet C wants nothing to do with that complicated brick talk and rhymes again and sticks with her methods.

8 x 9 = 80 – 8 = 72

8 x 10 = 8 0 = 80

8 x 11 = 8 8 = 88

8 x 12 = (8 x 10) + (8 x 2) = 80 + 16 = 96

Let's summarize this chapter that combines double fours, rhymes, and more.

8 x 1 = 4 x 1 = 4 (pause) x 2 = 8

8 x 2 = 4 x 2 = 8 (pause) x 2 = 16

8 x 3 = 4 x 3 = 12 (pause) x 2 = 24

8 x 4 = 4 x 4 = 16 (pause) x 2 = 32

(pause)

8 x 5 = 4 x 5 = 20 x 2 = 40

8 x 6 = 6 x 8, Beau got a play date, so 6 x 8 = 48.

8 x 7 = 7 x 8, Beau eats well and doesn't lick bricks, 7 x 8 = 56.

8 x 8, Beau ate and ate, got sick on the floor, 8 x 8 is sixty-four.

(pause)

8 x 9 = 80 – 8 = 72

8 x 10 = 8 0 = 80

8 x 11 = 8 8 = 88

8 x 12 = (8 x 10) + (8 x 2) = 80 + 16 = 96

If you are a kinesthetic learner who likes to do things with their hands, I think you'll like the next chapter on nines.

Chapter 9 – 9s, The Fingers Trick and More

In previous chapters we numbered each of our fingers in this way.

1. Above our left pinkie or little finger, we'll put a one.
2. Above our left ring finger, we'll put a two.
3. Above our left middle or long finger, we'll put a three.
4. Above our left index or pointer finger, we'll put a four.
5. Above our left thumb, we'll put a five.

Now we'll do the right hand.

6. Above our right thumb, we'll put a six.
7. Above our right index finger, we'll put a seven.
8. Above our right middle finger, we'll put an eight.
9. Above our right ring finger, we'll put a nine.
10. Above our right pinkie or little finger, we'll put a ten.

Let's use 9 x 7 as our example. We will push down our 7th finger, the right index finger, like typing a letter on a keyboard.

To the left of that 7th finger are the five fingers on your left hand and the right thumb on your right hand for a total of six fingers to the left of that 7th finger.

On the right of that seventh finger are your right middle finger, right ring finger, and right pinkie finger for a total of three fingers to the right of that seventh finger.

With six fingers on the left and three on the right, we find the solution to 9 x 7 = 6 3 or 63. This fingers technique works for numbers 1 through 10.

Triplet A – 9s, Multiply by 10 and Subtract and the Fingers Trick – 1 to 4

Triplet A starts off by using Triplet C's trick from many chapters where she multiplies by 10 and then subtracts the number.

9 x 1 = 10 x 1 = 10 (pause) – 1 = 9

9 x 2 = 10 x 2 = 20 (pause) – 2 = 18

9 x 3 = 10 x 3 = 30 (pause) – 3 = 27

9 x 4 = 10 x 4 = 40 (pause) – 4 = 36

But finds the finger trick a little faster where:

9 x 1 = 0 fingers left, 9 fingers right, 0, 9 = 9

9 x 2 = 1 finger left, 8 fingers right, 1, 8 = 18

9 x 3 = 2 fingers left, 7 fingers right, 2, 7 = 27

9 x 4 = 3 fingers left, 6 fingers right, 3, 6 = 36

Triplet B – 9s, Multiply by 10 and Subtract and the Fingers Trick – 5 to 8

Triplet B also gives both techniques a try. First, she uses Triplet C's trick of multiplying by 10 and then subtracting the number.

9 x 5 = 10 x 5 = 50 (pause) – 5 = 45

9 x 6 = 10 x 6 = 60 (pause) – 6 = 54

9 x 7 = 10 x 7 = 70 (pause) – 7 = 63

9 x 8 = 10 x 8 = 80 (pause) – 8 = 72

But Triplet B also finds the finger trick a little faster where:

9 x 5 = 4 fingers left, 5 fingers right, 4, 5 = 45

9 x 6 = 5 fingers left, 4 fingers right, 5, 4 = 54

9 x 7 = 6 fingers left, 3 fingers right, 6, 3 = 63

9 x 8 = 7 fingers left, 2 fingers right, 7, 2 = 72

However, we'll have to see what happens with Triplet C as there are only 10 fingers. Now, there is a primate that lives in the forest that has 12 fingers, it's called the aye-aye, spelled a-y-e, hyphen, a-y-e that is from Madagascar. It's got a long thin fingers and large eyes and ears in the middle of a round head.

Triplet C – 9s, Multiply by 10 and Subtract and the Fingers Trick – 9 to 12

Triplet C is pretty happy with her methods because they work so quickly.

9 x 9 = 10 x 9 = 90 (pause) – 9 = 81

9 x 10 = 9 0, 90

9 x 11 = 9 9, 99

9 x 12 = (9 x 10) + (9 x 2) = 90 + 18 = 108

Still, she gives the finger trick a try.

9 x 9 = 8 fingers left, 1 finger right, 8, 1 = 81

9 x 10 = 9 fingers left, 0 fingers right, 9, 0 = 90

The problem is, what happens with 9 x 11? Well, let's see if it would work with the aye-aye. If you had 12 fingers and put the 11th finger down, you would have 10 fingers to the left and one finger to the right, so it doesn't work out. Looks like for Triplet C, she's going to stay with the method that's been working all along. Let's review to make sure we have these down.

9 x 1 = 0 fingers left, 9 fingers right, 0, 9 = 9

9 x 2 = 1 finger left, 8 fingers right, 1, 8 = 18

9 x 3 = 2 fingers left, 7 fingers right, 2, 7 = 27

9 x 4 = 3 fingers left, 6 fingers right, 3, 6 = 36

(pause)

9 x 5 = 4 fingers left, 5 fingers right, 4, 5 = 45

9 x 6 = 5 fingers left, 4 fingers right, 5, 4 = 54

9 x 7 = 6 fingers left, 3 fingers right, 6, 3 = 63

9 x 8 = 7 fingers left, 2 fingers right, 7, 2 = 72

(pause)

9 x 9 = 10 x 9 = 90 (pause) − 9 = 81

9 x 10 = 9 0, 90

9 x 11 = 9 9, 99

9 x 12 = 9 x 10 + 9 x 2 = 90 + 18 = 108

We know the next chapter is going to be easy. We've already multiplied nine numbers times ten and with the ten rule, it's an easy day.

Chapter 10 – 10s, Add a Zero

This chapter, much like Chapter 1 allows us to use a rule that makes it straightforward in our calculations.

Triplet A – 10s, Add a Zero - 1 to 4

Triplet A starts off easy enough by adding zeros.

10 x 1 = 1 0, 10

10 x 2 = 2 0, 20

10 x 3 = 3 0, 30

10 x 4 = 4 0, 40

Triplet B – 10s, Add a Zero – 5 to 8

Triplet B is done quickly too.

10 x 5 = 5 0, 50

10 x 6 = 6 0, 60

10 x 7 = 7 0, 70

10 x 8 = 8 0, 80

Triplet C – 10s, Add a Zero – 9 to 12

Triplet C is just as happy with a single technique.

10 x 9 = 9 0, 90

10 x 10 = 10 0, 100

10 x 11 = 11 0, 110

10 x 12 = 12 0, 120

Let's review for consistency's sake, but a nice quick chapter.

10 x 1 = 1 0, 10

10 x 2 = 2 0, 20

10 x 3 = 3 0, 30

10 x 4 = 4 0, 40

(pause)

10 x 5 = 5 0, 50

10 x 6 = 6 0, 60

10 x 7 = 7 0, 70

10 x 8 = 8 0, 80

(pause)

10 x 9 = 9 0, 90

10 x 10 = 10 0, 100

10 x 11 = 11 0, 110

10 x 12 = 12 0, 120

Now, with our twins' rule, Triplet A and Triplet B will not have trouble with the 11s, but Triplet C will have to be clever.

CHAPTER 11 – 11s, TWINS AND SPLIT, ADD, INSERT (SPAIN MNEMONIC)

For the numbers 1 through 9, it's straightforward what we need to do using our twins' rule. When we multiply a number from 1 to 9 times 11, we get that number twice. For example, 8 x 11 = 8 8, 88. However, that rule doesn't work with 10, 11, and 12, so Triplet C will come up with a new method.

TRIPLET A – 11s, TWINS - 1 TO 4

Again, Triplet A starts off easy enough by using the number twice when multiplying by 11.

11 x 1 = 1 1, 11

11 x 2 = 2 2, 22

11 x 3 = 3 3, 33

11 x 4 = 4 4, 44

TRIPLET B – 11s, TWINS – 5 TO 8

Triplet B also says thank you very much and does the same.

11 x 5 = 5 5, 55

11 x 6 = 6 6, 66

11 x 7 = 7 7, 77

11 x 8 = 8 8, 88

TRIPLET C – TWINS, ADD A ZERO, SPLIT, ADD, AND INSERT (SPAIN MNEMONIC)

Triplet C can use the twins' rule for eleven times nine, so

11 x 9 = 9 9, 99

She can use the add a zero technique to figure out eleven times ten.

11 x 10 = 11 0, 110

However, what will she do with 11 x 11?

When you multiply a number greater than 10 times 11, you can use the SPAIN mnemonic to do this trick.

First, you SPLIT the numbers 1 and 1 and leave a space in between.

Second, you ADD the two numbers, in this case 1 + 1 = 2.

Third, you INSERT that number 2 between the 1 and 1 to get 1, 2, 1 or 121.

To make it more visual, I like to pretend the two ones in the number eleven are doors and and separate them by putting a space in between them.

So, SPLIT one space one with an underscore (1_1).

Then ADD the two numbers, one plus one equals two (1 + 1 = 2).

INSERT that number two in the middle of the doors to get one, two, one or one-hundred and twenty-one (1, 2, 1 or 121).

The reason I call it the SPAIN mnemonic is that the word "split" starts with the letters "s-p," the word "add" starts with a letter "a," and the word "insert" starts with an "i-n." So, you remember to split, add, and insert by remembering Sp-a-in, Spain.

Let's try this method again with the number 12.

First, you SPLIT the numbers 1 and 2 and leave a space in between.

Second, you ADD the two numbers, in this case 1 + 2 = 3.

Third, you INSERT the "3" between the 1 and 2 to get 1, 3, 2 or 132.

Does this work for big two-digit numbers? Let's try a big two-digit number 98.

First, you SPLIT the numbers 9 and 8 and leave a space in between.

Second, you ADD the two numbers, in this case 9 + 8 = 17

Third, you INSERT the "17" between the 9 and 8 to get 9, 1, 7, 8, 9,178 which is unfortunately wrong.

What you do in this case is you keep the second digit, 7 from the seventeen and INSERT it in the middle and ADD the 1 to the first 9 to get 10.

Then you have 10, 7, 8 or 1078 which is the correct answer. Let's get back to summarizing how we'll tackle 11 times 9, 10, 11, and 12.

11 x 9 = 9 9, 99

11 x 10 = 11 0, 110

11 x 11 = 1, 1+1, 1 or 1, 2, 1 or 121

11 x 12 = 1, 1+2, 2 or 1, 3, 2 or 132.

Let's review all of what we've learned in this chapter.

11 x 1 = 1 1, 11

11 x 2 = 2 2, 22

11 x 3 = 3 3, 33

11 x 4 = 4 4, 44

(pause)

11 x 5 = 5 5, 55

11 x 6 = 6, 6, 66

11 x 7 = 7 7, 77

11 x 8 = 8 8, 88

(pause)

11 x 9 = 9 9, 99

11 x 10 = 11 0, 110

11 x 11 = 1, 1+1, 1 or 1, 2, 1 or 121

11 x 12 = 1, 1+2, 2 or 1, 3, 2 or 132.

Now, what are we going to do for the 12s?

Well, when one triplet brings home an Acai bowl or mozzarella sticks, you know the other triplets are going to take a little something for themselves. In the same way, Triplet A and Triplet B will take Triplet C's method for multiplying by 12.

Chapter 12 – 12s, Times 10, Times 2 and Add Them

Although in this chapter we're going to use the multiply by 10 and multiply by 2 then add rule, we'll also mention much easier ways to get the answer when it makes sense like with the 1s and 10s.

Triplet A – 12s, Times 10, times 2 and Add Them - 1 to 4

Triplet A starts off using the rule that Triplet C has been using all this time.

12 x 1 = (10 x 1) + (2 x 1) = 10 + 2 = 12, but 12 x 1 = 12 is easier.

12 x 2 = (10 x 2) + (2 x 2) = 20 + 4 = 24, but 12 + 12 is easier.

12 x 3 = (10 x 3) + (2 x 3) = 30 + 6 = 36

12 x 4 = (10 x 4) + (2 x 4) = 40 + 8 = 48

Triplet B – 12s, Times 10, times 2 and Add Them - 5 to 8

Triplet B will do the same.

12 x 5 = (10 x 5) + (2 x 5) = 50 + 10 = 60

12 x 6 = (10 x 6) + (2 x 6) = 60 + 12 = 72

12 x 7 = (10 x 7) + (2 x 7) = 70 + 14 = 84

12 x 8 = (10 x 8) + (2 x 8) = 80 + 16 = 96

Triplet C– 12s, Times 10, times 2 and Add Them - 9 to 12

Triplet C will remind her sisters she showed them how to do this, then use that technique as well.

12 x 9 = (10 x 9) + (2 x 9) = 90 + 18 = 108

12 x 10 = (10 x 10) + (2 x 10) = 100 + 20 = 120, but 12 x 10 = 12 0, 120 is an easier method.

12 x 11 = (10 x 11) + (2 x 11) = 110 + 22 = 132

I'm not sure if that method easier than our (SPAIN mnemonic) Split, Add, Insert 11 x 12 = 1, 1+2, 2 = 1, 3, 2 = 132 from the last chapter

12 x 12 = (10 x 12) + (2 x 12) = 120 + 24 = 144

Here's the review of this chapter, but I want you to notice the pattern in the ones place moving from 2, 4, 6, 8, to 0.

12 x 1 = (10 x 1) + (2 x 1) = 10 + 2 = 12

12 x 2 = (10 x 2) + (2 x 2) = 20 + 4 = 24

12 x 3 = (10 x 3) + (2 x 3) = 30 + 6 = 36

12 x 4 = (10 x 4) + (2 x 4) = 40 + 8 = 48

(pause)

12 x 5 = (10 x 5) + (2 x 5) = 50 + 10 = 60

12 x 6 = (10 x 6) + (? x 6) = 60 + 12 = 72

12 x 7 = (10 x 7) + (2 x 7) = 70 + 14 = 84

12 x 8 = (10 x 8) + (2 x 8) = 80 + 16 = 96

(pause)

$12 \times 9 = (10 \times 9) + (2 \times 9) = 90 + 18 = 108$

$12 \times 10 = (10 \times 10) + (2 \times 10) = 100 + 20 = 120$

$12 \times 11 = (10 \times 11) + (2 \times 11) = 110 + 22 = 132$

$12 \times 12 = (10 \times 12) + (2 \times 12) = 120 + 24 = 144$

To help you remember you can make yourself aware of a pattern. The last digit goes 2, 4, 6, 8, 0 as the answers from 12, 24, 36, 48, and 60. That keeps on going forever.

You can multiply 12 by 13, 14, and 15 to double check.

$12 \times 13 = (10 \times 13) + (2 \times 13) = 130 + 26 = 156$

$12 \times 14 = (10 \times 14) + (2 \times 14) = 140 + 28 = 168$

$12 \times 15 = (10 \times 15) + (2 \times 15) = 150 + 30 = 180$

You can see that when you multiply 12 times 11, 12, 13, 14, and 15, the last ones digits end up again as 2, 4, 6, 8, 0 in 132, 144, 156, 168 and 180. Although we've done the times table to 12, what I want to do is give you an easier way to remember the doubles such as 5×5, 6×6 and so on, in Chapter 13.

CHAPTER 13 – MULTIPLYING DOUBLES WITH RHYMES

As we've seen, it's better if you have more than one way to remember a math fact. Rhymes work well to help us remember the doubles such as one times one, two times two, all the way to twelve times twelve (1 x 1, 2 x 2 all the way to 12 x 12).

First, you need two words that rhyme with the numbers.

Second, the *rhyme* needs to create an image in your *mind*.

TRIPLET A – DOUBLES RHYMES - 1 TO 4

1 x 1 is still 1, so you're already *done*, 1 x 1 is only *one*.

2 x 2, 2 pairs of shoes are 2 shoes and 2 *more*, that equals *four*.

3 x 3, skis don't hit the *pine*, 3 x 3 is only *nine*.

With the threes, you could think about two words that rhyme with three, like skis and trees. A pine is a type of tree, so you can imagine a skier going through the trees and telling them to watch out for a pine tree which comes out like this: Three times three, skis don't hit the pine, three times three is only nine.

4 x 4, Four by fours are snow *machines*, you can drive one when you turn 16, so 4 x 4 is *sixteen*.

A four by four is a truck that has the same amount of power going to all four wheels, unlike a front wheel drive car which just has the power going to the front wheels. In Iowa, when it snows, it's better to have a four by four. You get your driver's license when you're 16 years old Let's combine these: Four times four, four by fours are snow machines, can drive one when you turn sixteen, so four times four is sixteen.

TRIPLET B – DOUBLES RHYMES - 5 TO 8

5 x 5, plenty of *dives*, 5 x 5 is *twenty-five*.

If you are on the diving team and dive each day of the week from Monday through Friday, that's five dives. If you do that for five weeks, that's twenty-five. Five times five, plenty of dives, five times five is twenty-five.

6 x 6, Beau chews dirty *sticks*, so 6 x 6 is *thirty-six*.

Beau, again spelled, B-e-a-u, is our green-eyed Vizsla, a type of dog pointer dog with a copper coat. A beau is a fashionable young man, as is our dog, Beau. So, you are going to picture Beau picking up a muddy stick in the yard, and just as puppies will do, he'll chew at it. So, this is our rhyme. Six times six, Beau chews dirty sticks, so six times six is thirty-six.

7 x 7, have *four* straight *lines*, so 7 x 7 is *forty-nine*.

With sevens, it was a bit tougher to remember this because not a lot of things rhyme with seven except the word heaven, so we can look at the seven and see that there are four straight lines that make up the two numbers. Then we have the picture in our brain by using this rhyme. Seven times seven have four straight lines, so seven times seven is forty-nine.

8 x 8, Beau ate and ate, got *sick* on the *floor*, so 8 x 8 is *sixty-four*.

Sometimes puppies like Beau get excited and hungry and eat and eat, and unfortunately, they might get a little bit sick and throw up on the floor. Yeah, it's a little gross, but that image gives us our next memorable rhyme. We're going to use the two homophones ate, a-t-e and eight, e-i-g-h-t. Remember, homophones are words that *sound* the same, the "*phone*" part, but are, in this case, spelled differently. Here's our rhyme again. Eight times eight, Beau ate and ate, got sick on the floor, so eight times eight is sixty-four.

TRIPLET C – DOUBLES RHYMES - 9 TO 12

9 x 9, to lose weight dad *runs*, so 9 x 9 is *eighty-one*.

I'm a dad, so a lot of times I like to run to try to lose weight and get in shape for a race. So, I came up with this complete rhyme. Nine times nine, to lose weight dad runs, nine times nine is eighty-one.

10 x 10, what is the answer I *wondered*, so 10 x 10 is *one-hundred*.

There are just not a lot of things that rhyme with a hundred, so wondered was my best bet and it worked out that Ten times ten, what is the answer I wondered, ten times ten is one hundred.

11 x 11, that makes one two *one*, so 11 x 11 is one-hundred *twenty-one*.

I was kind of running out of rhyme energy here, if you've got a better one, let me know. For now it's eleven times eleven, that makes one two one, so eleven times eleven is one twenty-one.

12 x 12, that makes one tied *score*, so 12 x 12 is one-hundred-forty-four.

Imagine on a scoreboard that the game is in the first period, one, the score is tied, four to four and that number twelve on one team is shooting against number twelve on the other team in soccer, basketball, or whatever sport matters most to you. Twelve times twelve, that makes one tied score, so twelve times twelve is one-hundred forty-four.

As with past chapters, it's good to review all that we've gone over.

1 x 1 is still 1, so you're already *done*, 1 x 1 is only *one*.

2 x 2, 2 pairs of shoes are 2 shoes and 2 *more*, that equals *four*.

3 x 3, skis don't hit the *pine*, 3 x 3 is only *nine*.

4 x 4, four by fours are snow *machines*, you can drive one when you turn 16, so 4 x 4 is *sixteen*.

(pause)

5 x 5, plenty of *dives*, 5 x 5 is *twenty-five*.

6 x 6, Beau chews dirty *sticks*, so 6 x 6 is *thirty-six*.

7 x 7, have *four* straight *lines*, so 7 x 7 is *forty-nine*.

8 x 8, Beau ate and ate, got *sick* on the *floor*, so 8 x 8 is *sixty-four*.

(pause)

9 x 9, to lose weight dad *runs*, so 9 x 9 is *eighty-one*.

10 x 10, what is the answer I *wondered*, so 10 x 10 is *one hundred*.

11 x 11, that makes one two *one*, so 11 x 11 is one-hundred *twenty-one*.

12 x 12, that makes one tied *score*, so 12 x 12 is one-hundred-forty-four.

SHORTENING THE RHYMES

While I think many of those rhyme are clever, it's a lot to remember the whole rhyme, so what we really want to do is to cut the rhymes to their essence so it acts as a quick mnemonic for us to remember them. We'll group them into fours because your short-term memory gets overwhelmed when you do more than that, so that it's easier to remember and then shorten them one more time.

1 x 1 = done, 1
2 x 2 = more, 4
3 x 3 = pine, 9
4 x 4 = machines, 16

Pause.

5 x 5 = dives, 25
6 x 6 = dirty sticks, 36
7 x 7 = four lines, 49
8 x 8 = sick floor, 64

Pause.

9 x 9 = lose weight runs, 81
10 x 10 = wondered, 100
11 x 11 = one two one, 121
12 x 12 = one tied score, 144

And that's it, hopefully, you are a big step closer to knowing your multiplication facts.

Afterword

I really didn't like math when I was in school and my first year in high school was even worse. I made it to a higher-level math in a college Calculus I class I took while in high school, but what made me appreciate math later in life was something I didn't expect.

As a grade school and high school student, a lot of the answers are very concrete in the classroom. However, outside of the classroom, as you take on leadership and family responsibilities a lot of times there is a gray area.

I don't know if I'm doing the right thing as a parent a lot of times, but I am darn sure that seven times eight equals fifty-six ($7 \times 8 = 56$). As I taught dosage calculations in pharmacology and a similar type of math, dimensional analysis in inorganic chemistry class, I felt such a calm knowing that yes, there is a single correct answer that I can be confident in.

So, while my math road was bumpy, I hope that later in life, as you run into problems that don't necessarily have correct answers that you come back to math as a place where you can be confident in a single, correct answer.